Animals in Antarctica

Julie Murray

Abdo Kids Junior
is an Imprint of Abdo Kids
abdobooks.com

Abdo
ANIMAL HABITATS
Kids

abdobooks.com

Published by Abdo Kids, a division of ABDO, P.O. Box 398166, Minneapolis, Minnesota 55439.
Copyright © 2021 by Abdo Consulting Group, Inc. International copyrights reserved in all countries.
No part of this book may be reproduced in any form without written permission from the publisher.
Abdo Kids Junior™ is a trademark and logo of Abdo Kids.

Printed in China

052020

092020

THIS BOOK CONTAINS
RECYCLED MATERIALS

Photo Credits: iStock, Shutterstock

Production Contributors: Teddy Borth, Jennie Forsberg, Grace Hansen

Design Contributors: Candice Keimig, Pakou Moua, Dorothy Toth

Library of Congress Control Number: 2019955548

Publisher's Cataloging-in-Publication Data

Names: Murray, Julie, author.

Title: Animals in Antarctica / by Julie Murray

Description: Minneapolis, Minnesota : Abdo Kids, 2021 | Series: Animal habitats | Includes online resources
    and index.

Identifiers: ISBN 9781098202071 (lib. bdg.) | ISBN 9781098203054 (ebook) | ISBN 9781098203542
    (Read-to-Me ebook)

Subjects: LCSH: Animals--Habitations--Juvenile literature. | Habitat (Ecology)--Juvenile literature. |
    Antarctica--Juvenile literature.

Classification: DDC 591.52--dc23

Table of Contents

Animals in Antarctica

Antarctica is a cold place. It is also called the South Pole.

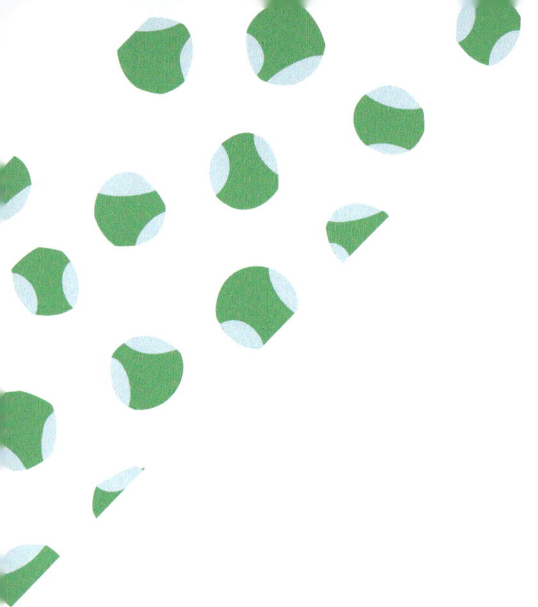

It is too cold for most animals.

Only a few live here.

Krill are in the ocean. They are food for some animals.

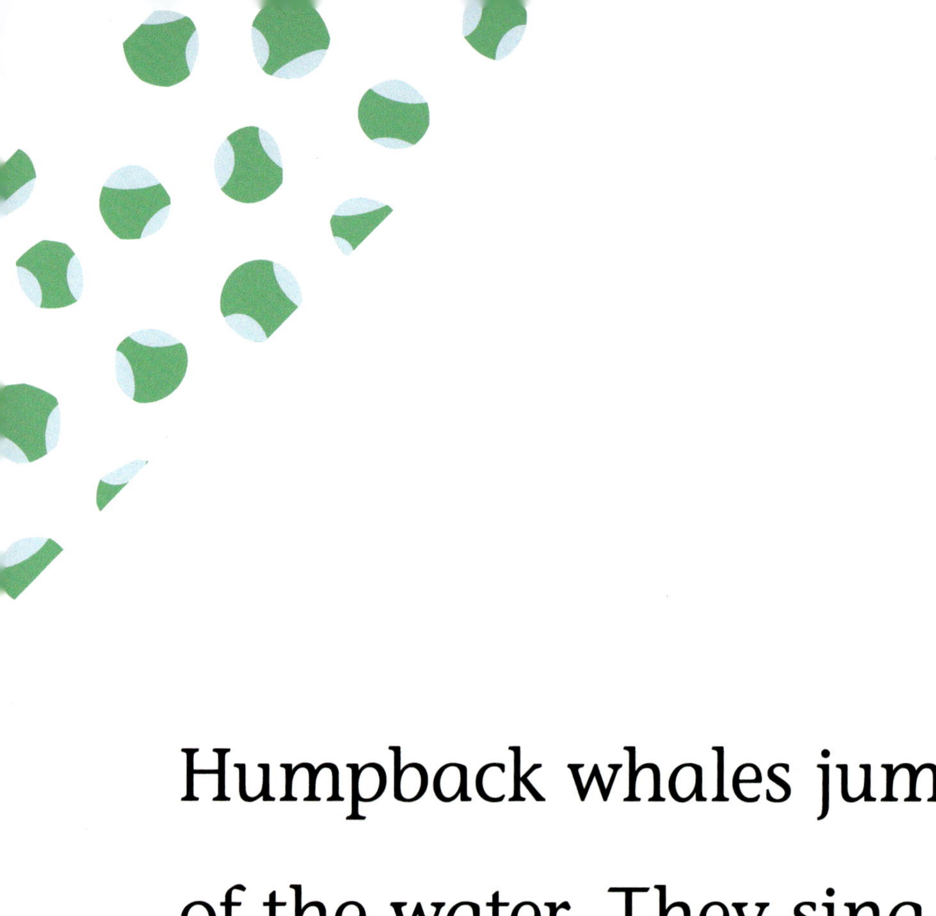

Humpback whales jump out
of the water. They sing too!

11

Fur seals have flippers.

They can move on land.

Penguins **huddle** together.

This keeps them warm.

emperor penguins
15

Orca whales live in pods.

Pods are groups.

Birds fly in the air.

They eat fish.

19

Blue whales are big. They are
the biggest animal on Earth!

More Animals in Antarctica

Adélie penguin

leopard seal

snow petrel

south polar skua

Glossary

huddle
to move close together. Penguins do this to share heat.

krill
a small shrimplike animal of the open seas.

Index

Visit **abdokids.com** to access crafts, games, videos, and more!